Yellowstone's Ecosystem

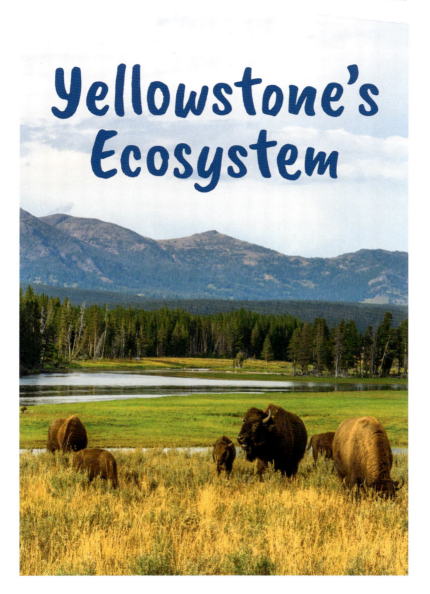

Caroline Tung Richmond

Grand Canyon of the Yellowstone

Consultants

Darrin Lunde
Collection Manager
National Museum of Natural History

Cheryl Lane, M.Ed.
Seventh Grade Science Teacher
Chino Valley Unified School District

Michelle Wertman, M.S.Ed.
Literacy Specialist
New York City Public Schools

Publishing Credits

Rachelle Cracchiolo, M.S.Ed., *Publisher*
Emily R. Smith, M.A.Ed., *SVP of Content Development*
Véronique Bos, *VP of Creative*
Dani Neiley, *Editor*
Robin Erickson, *Senior Art Director*

Smithsonian Enterprises

Avery Naughton, *Licensing Coordinator*
Paige Towler, *Editorial Lead*
Jill Corcoran, *Senior Director, Licensed Publishing*
Brigid Ferraro, *Vice President of New Business and Licensing*
Carol LeBlanc, *President*

Image Credits: p.13 Darrah Leffler; p.16 Carsten Steger; p.17 Nicole R. Fuller/Science Source Images; p.18 National Park Service/ incidencematrix; p.22 Alamy Stock Photo; all other images from iStock and/or Shutterstock, or in the public domain.

Library of Congress Cataloging in Publication Control Number: 2024024234

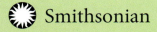

© 2025 Smithsonian Institution. The name "Smithsonian" and the Smithsonian logo are registered trademarks owned by the Smithsonian Institution.

This book may not be reproduced or distributed in any way without prior written consent from the publisher.

5482 Argosy Avenue
Huntington Beach, CA 92649
www.tcmpub.com
ISBN 979-8-7659-6861-1
© 2025 Teacher Created Materials, Inc.
Printed by: 51497
Printed in : China

Table of Contents

Life in Abundance 4

Examining the Ecosystem 6

Life Under a Microscope 16

Effects of Climate Change 20

Protected Lands 26

STEAM Challenge 28

Glossary 30

Index 31

Career Advice 32

Life in Abundance

An elk grazes in the mountains, its ears pricked for the sounds of an approaching gray wolf. To the east, a herd of bison grazes in a sunny meadow. A single deer laps water from a stream as an eagle swoops down from the sky, snatching a marmot in its claws. It soars through the air, flying back to its nest of hungry eaglets. And to the south, a colorful hot spring bubbles with near-boiling water. Surprisingly, life can exist even in extreme temperatures. Tiny creatures known as **thermophiles** call hot springs and **geysers** home.

Yellowstone National Park is one of the most popular national parks in the United States, and it's not hard to see why. This amazing wildlife and more can be found there. And the park's figure-eight-shaped road takes visitors past many astonishing sights. These include tall waterfalls, wide-open meadows, and ancient rock formations.

Yellowstone is part of one of the most unique **ecosystems** in the world. This is all thanks to the thousands of plant and animal species that flourish across the land. Life thrives there, but death has a role, too. Learning about the ecosystem, thermal features, and the effects of **climate change** at the park unlocks a unique view into this special part of the United States.

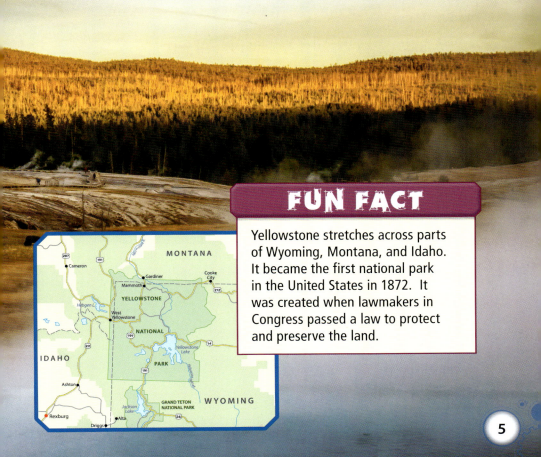

FUN FACT

Yellowstone stretches across parts of Wyoming, Montana, and Idaho. It became the first national park in the United States in 1872. It was created when lawmakers in Congress passed a law to protect and preserve the land.

Examining the Ecosystem

Yellowstone National Park is part of one of the biggest ecosystems on Earth. It is at the heart of the Greater Yellowstone Ecosystem. This area covers a large amount of land in Wyoming, Montana, and Idaho. Yellowstone's boundaries mark just a small part of this greater ecosystem. But the park's unique protections provide an important space for wildlife and plants. Here, scientists have countless opportunities to study how life thrives.

Yellowstone's ecosystem includes all the animals, plants, and **microbes** that live there. It also includes the nonliving things in the park. Soil, rocks, and bodies of water are part of the ecosystem, too. Ultimately, the living and nonliving parts of Yellowstone's ecosystem are linked together. They are connected through cycles of energy and nutrients.

ARTS

Painting the Park

In 1871, a painter named Thomas Moran set out on a 40-day journey. He traveled in the Wyoming Territory and painted what he saw. In the end, he had numerous paintings of the area's geysers, lakes, and mountains. His artwork helped convince lawmakers to create Yellowstone National Park.

All plants and animals in an ecosystem need energy to live. These living things can be sorted into three groups: producers, consumers, and decomposers. The way that energy flows through these three groups can make up a **food chain**. Each group plays a specific role in Yellowstone's ecosystem. Producers, such as plants, use energy from the sun to make their own food. Consumers eat these plants and other animals. Decomposers break down dead plants and animals. Their nutrients go into the soil, where new plants can sprout. All together, these groups bring balance to the park's ecosystem.

Plants like this lodgepole pine tree are producers.

Grizzly bears are consumers.

Fungi, including these meadow mushrooms, are a type of decomposer.

7

Producers: The Chefs

Producers are at the first level of every food chain. At Yellowstone, they include grasses, shrubs, and trees. Producers are a bit like chefs: they make their own food. They take energy from the sun and use it to create food through **photosynthesis**. This process sustains them throughout their lives. Eventually, they become food for consumers and decomposers. Producers form the backbone of the park's ecosystem, passing their energy up a food chain.

In Yellowstone's forests, producers include trees that cover 80 percent of the land. Willow and fir trees nourish moose who love to munch on leaves. Seeds from conifer trees sustain red squirrels as they scurry through the woods. And during the winter, tree bark can help snowshoe hares survive the coldest months.

Birds, foxes, and raccoons like to eat Oregon grape plants that grow along Yellowstone Lake.

Moose eat 18–27 kilograms (40–60 pounds) of plant material a day!

Yellowstone has many wetlands throughout the park.

Yellowstone's rivers and lakes contain many aquatic plants and algae. These producers feed all sorts of animals, from the tiniest tadpoles to 318-kilogram (700-pound) elk. Algae is a common food for flies and trout. In ponds, water lilies are both pretty and practical. Beavers eat them when land-based plants are scarce.

Grasses and **sedges** are key producers in Yellowstone's valleys. Bison **migrate** across the park to nibble on these sprouting plants. Elk and bighorn sheep also rely heavily on grasses for their meals. Small mammals, such as voles and marmots, get their nutrients from grasses, too.

Consumers: Hungry Creatures

Consumers come in all shapes and sizes at Yellowstone. There are three main types of consumers: primary, secondary, and tertiary. Some of them eat only plants, some eat other animals, and some eat a mix of both. Energy passes up a food chain through these consumers. Ultimately, they all become food for decomposers.

Primary consumers get their energy by eating producers. On land, these consumers include furry pikas that chomp on sedges and lichen. Bison that graze on grasses are consumers, too. In the water, consumers come in the form of **suckers** and trout that feed on lake algae. Consumers at this level help control the growth of plants at the park. They also provide energy for other consumers.

Yellowstone cutthroat trout

To survive the winter, pikas store food inside the rock piles where they live.

Secondary consumers mostly eat primary consumers and producers. One example is the weasel-like marten. Martens hunt small consumers, such as mice and squirrels. They also forage for nuts and berries that are made by producers.

Martens climb trees to gain access to nuts and berries.

Tertiary consumers sit at the top of a food chain. They are known as **predators** because they mostly eat primary and secondary consumers. In Yellowstone, they include grizzly bears and gray wolves. These animals prey on others to survive. And they don't worry about being eaten themselves!

Grizzly bears prey on bison and elk.

Gray Wolf Numbers

In 1926, there were zero gray wolves in Yellowstone. People had been allowed to hunt them because they were seen as vicious predators. But when these wolves disappeared, the park's ecosystem became unbalanced. By hunting elk and bison, wolves helped prevent **overgrazing**. Over time, scientists realized that they needed to bring the wolves back. In the 1990s, gray wolves were reintroduced to the park. In 2023, 108 gray wolves were counted in the park.

Decomposers: Clean-Up Crew

When a producer or a consumer dies, their energy passes on to decomposers. Decomposers complete a food chain. They break down the bodies and tissues of dead plants and animals. While their role in a food chain might seem minor, they allow life in Yellowstone to keep thriving.

Most decomposers are bacteria that are too tiny for people to see without a microscope. Bacteria live in the soil and water at the park, feeding off the remains of plants and animals. They produce **enzymes** that transform dead **organic** matter into nutrients. These nutrients are then returned to the land, lakes, and rivers. This recycling of nutrients helps new producers grow.

puffball mushrooms

Some fungi, such as mushrooms and molds, are decomposers you can easily see. For instance, puffballs are mushrooms that can grow as big as beach balls! They break down the remains of decaying worms and plants. This helps them grow. They return nutrients to the soil, too.

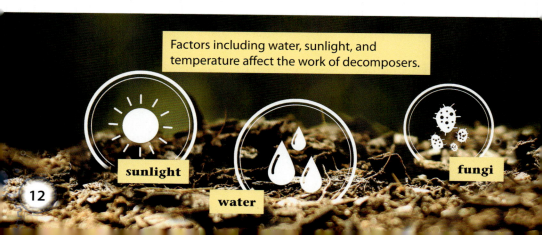

Factors including water, sunlight, and temperature affect the work of decomposers.

sunlight

water

fungi

Scavengers

Some animals assist decomposers with their work. These animals are known as scavengers, and they eat **carrion**. When a scavenger has eaten their fill, decomposers come along to break down the remains further. Scavengers come in a range of sizes and species. Tiny insects called *carrion beetles* feast on remains. Large predators, such as grizzly bears, can be scavengers as well. They like to eat carrion after they wake up from their winter hibernation.

ridged carrion beetle

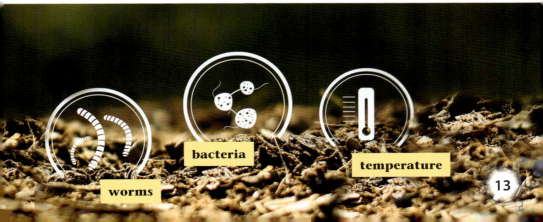

worms

bacteria

temperature

Let's Be Friends!

Some animals in Yellowstone's ecosystem have learned how to **coexist**. Working together, they help one another survive. These surprising examples show how species in the park collaborate.

Wolves and ravens have a special relationship at Yellowstone. When a wolf pack hunts elk, ravens are known to follow them. After the wolves make a kill, the ravens swoop in to eat the meat, too. They scavenge whatever they can get. And in return, ravens act as a security system for the wolves. They croak out in warning if another predator comes too close.

A gray wolf and ravens feast on a bison carcass.

raven

SCIENCE

Research at Yellowstone

Yellowstone has an extensive research program. Every year, hundreds of scientists work in the park on a range of projects. Some scientists have studied relationships among animals. Others have studied tree rings or found new microbes. Before they can start working, most scientists have to apply for research permits. They also must follow rules to protect the ecosystem.

Bison and dung beetles have a special bond at the park, too. As you might have guessed from their name, dung beetles have a unique connection to bison poop, or pies. These pies serve as food for dung beetles and other insects. Dung beetles also bury the pies underground, which helps break down the dung. As it breaks down, it provides nutrients for the soil. New grasses then grow from the soil, feeding the bison and starting this cycle all over again.

Some dung beetles roll dung into round balls.

Finding a friend can be especially helpful during the harsh winters. Elk, bison, and deer try to stay in the valleys, avoiding the high peaks in the park. These areas can make survival easier. As they migrate, they follow one another's tracks through the snow, saving their energy.

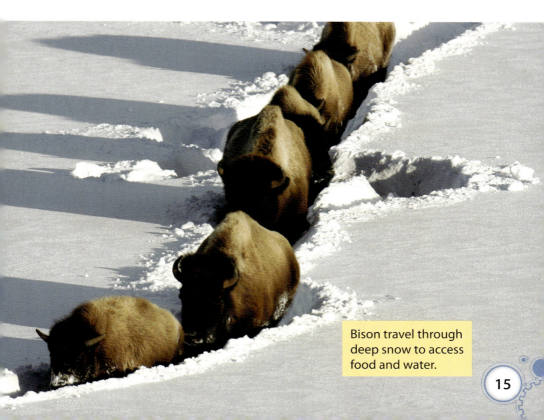

Bison travel through deep snow to access food and water.

Life Under a Microscope

While Yellowstone's stunning wildlife draws visitors to the park, its geysers and hot springs are just as famous. At first glance, these water features don't seem compatible with life. After all, geysers spew boiling water high into the air while hot springs can soar to 93 °C (199 °F). But take a closer look, and you'll find that life can prosper even under these scorching conditions.

Some microscopic organisms grow best in high temperatures. They are called *thermophiles*. When they group together, they form what looks like clumps of fuzzy carpet. These are called *mats*, and they can appear as different shades of colors.

Grand Prismatic Spring is Yellowstone's largest hot spring.

Under a microscope, hot spring water shows bacteria cells.

The thermophiles in Yellowstone's bubbling waters come from three main groups of organisms. First, archaea are tough, single-celled organisms. They do not have a nucleus or any other organelles. Their strong cell walls keep them safe from heat and acid. Second, bacteria are also single-celled organisms. Bacteria can be rod- or sphere-shaped. Together, they form strands that combine into mats. Third, there are eukarya. These include specific plants and animals that love the heat. Some of them are certain types of algae that can be seen in hot springs. Larger eukarya include flies that live among colorful mats of bacteria.

FUN FACT

Some of Yellowstone's geysers and hot springs produce steam—and gas. This gas is the reason some areas in the park smell like rotten eggs! The stinky culprit is called *hydrogen sulfide*. It's a chemical compound that creates a colorless, smelly gas.

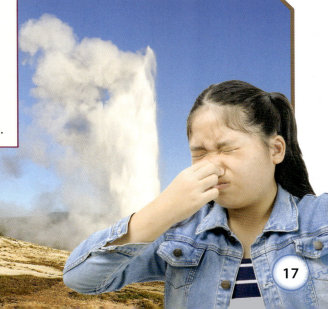

Gazing into Geysers

Did you know that Yellowstone is home to nearly half of the world's geysers? Most of them can be found in an area known as the Upper Geyser Basin. Five of the park's major geysers can be found there, including one of the most famous—Old Faithful. Overall, this area has hundreds of hydrothermal features. Let's dive into some of them!

Pump Geyser

Pump Geyser may be small, but its water flow is nearly constant. This geyser frequently shoots water up into the air. The water then runs down a hill and creates small channels, a welcoming place for thermophiles. Lots of bacteria live there in mats. In some places, they look like streamers, flowing in the water as if they were waving.

Castle Geyser

Castle Geyser gets its name from its unique structure. It is made of fragile minerals that have been built up over thousands of years. This geyser has a predictable eruption schedule. Roughly every 14 hours, it erupts for about 20 minutes at a time. Then, it loudly bellows steam for about 30 to 40 minutes.

Pump Geyser

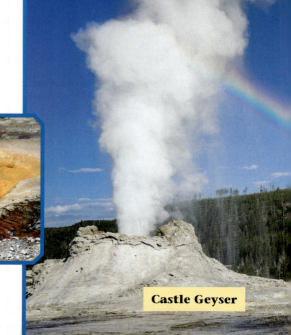

Castle Geyser

Old Faithful

Like Castle Geyser, Old Faithful erupts on a predictable schedule. On average, it erupts every 90 minutes. At its highest, water from this geyser shoots about 56 meters (184 feet) into the air! Visitors to the park can gather on a boardwalk to watch the show.

TECHNOLOGY

Mapping Geysers

With their tall explosions of water, geysers look impressive on the surface. But they have a lot going on underground, too. To map the areas beneath geysers, scientists use special tools called *seismometers*. These tools measure ground vibrations and movements. Scientists can use them to trace the directions of ground movements and then map the sources.

Old Faithful

Effects of Climate Change

Climate plays a vital role in the ecosystem at Yellowstone. Weather conditions have always affected life at the park. But now, climate change is speeding up across our planet. Over time, the park's average yearly temperatures have increased. This increase could have damaging effects across the park. It may cause harmful chain reactions that transform the land.

Due to climate change, weather events can become more extreme. One example of this can be seen in the increased wildfire season. With warmer average temperatures, there are more hot days. More hot days means a higher risk of wildfires. Plus, warmer temperatures can mean less rain and possible periods of drought. Less moisture can lead to drier forests and meadows. These conditions can cause larger wildfires. Dry trees and grasses serve as fuel.

Plant and animal species can be dangerously affected by climate change, too. In warm temperatures, **pests** such as mountain pine beetles can reproduce more quickly. These beetles are a huge threat to whitebark pine trees. They devour the bark and destroy the trees. In larger numbers, these beetles could damage more trees. Then, with fewer pine trees in the ecosystem, certain animals would struggle to find food. They would no longer have access to the seeds from the trees.

mountain pine beetle

The National Park Service estimates that Yellowstone experiences anywhere from 5 to 78 fires each year.

The Flood of 2022

On the morning of June 13, 2022, historic flooding occurred in northern Yellowstone. Some parts of the park got nearly 25 centimeters (10 inches) of **snowmelt** and rainfall. This large amount of unexpected water caused the Yellowstone River to flood. The flooding wrecked roads, destroyed homes, and ruptured a sewage line in the park. It also caused mudslides and altered river channels. It was one of the worst floods to ever happen in the park.

When the flooding stopped, some areas of the park looked very different. Debris including fallen trees and rocks had flowed downstream, cutting new paths for rivers. Hillsides and riverbanks had been eroded by the forceful water. Sections of the park's roads that led to other towns were impassable. And the local wildlife was affected, too. Bird nests were washed away by the rushing waters.

Sections of the park's North Entrance road were severely damaged in the flooding.

Rainbow and cutthroat trout declined in number. Luckily, larger animals, such as elk and bison, were able to move fast enough to avoid the flood.

In time, the park's staff was able to restore the broken roads and buildings across the park. But they now keep a close watch on the threat of future flooding. That's because climate change can make extreme rainfall worse in many storms. Plus, warmer temperatures from climate change increase snowmelt. When these two factors combine, flash floods can occur.

ENGINEERING

Road Repairs

High waters during the flood damaged one of the park's main entrance roads. But park crews worked quickly. Just two days after the flood, they hauled in more than 18,144 metric tons (20,000 tons) of gravel. They used the gravel to build a one-lane path for employee traffic. Full repairs took months, and the road was opened again to visitors on November 1, 2022.

Addressing Changes in Climate

Yellowstone's scientists work to address the effects of climate change on the park's ecosystem. Their first line of defense is research. The more information they can gather, the more they can understand what might happen and plan for next steps. That's why scientists in the park monitor wildlife and weather shifts every year. For example, they investigate how warmer streams affect fish. They track how lower amounts of rain affect wetland frogs. And they gather plant and soil samples every season to learn how plants respond to change. Armed with research data, Yellowstone's scientists decide how to best protect the land.

An archaeologist examines soil color to analyze conditions inside the park.

These are just a few actions the park's staff have taken. In lakes and streams, they also monitored fish populations closely. Then, they developed a **conservation** plan to increase populations. In addition, workers have installed renewable energy sources. One of the park's ranger stations is run by solar energy. And some buildings in the park run on **hydropower**. Finally, Yellowstone's **park rangers** have emphasized education. They teach tourists about the effects of climate change and what they can do to help. This work extends beyond the park's borders, too. Rangers and scientists at Yellowstone work with outside groups and universities. Together, they work on ecosystem preservation.

A scientist takes measurements in a creek to determine the water's speed.

Park rangers teach visitors about the ecosystem.

Protected Lands

For more than 150 years, Yellowstone has been a national treasure. Visitors have come from all over the world to glimpse its grizzly bears, gape at its geysers, and hike through its dense forests. With its striking landscapes and wildlife, there's no other place in the world quite like it.

The park's ecosystem has thrived throughout time, even long before the first Indigenous peoples set foot upon it. Plants, animals, fungi, and microbes have called it home for millions of years. Year after year, they live, grow, and die. But even in death, they remain part of an important cycle. As they break down into nutrients for the soil, they help new plant life to take root and grow.

Life will keep flourishing at Yellowstone thanks to protections from lawmakers. However, some changes in the ecosystem will occur due to climate change. The risk of wildfires and flash floods will increase. And as temperatures rise, plants and wildlife will be threatened. But the park's staff and scientists—along with its tourists and even readers like you—can do their parts to protect it.

Whether you live near or far from Yellowstone's ecosystem, you are still connected to it. That's because we are all part of the ecosystems that make up Earth.

Lower Falls of the Yellowstone River

FUN FACT

Yellowstone's name comes from the main river that flows through the park. The Hidatsa tribe had a name for the river in their language. It translated to "Rock Yellow River." French-Canadian explorers called it *Roche Jaune* in French. And an English explorer translated that to "Yellow Stone." Eventually, the name became one word.

STEAM CHALLENGE

Define the Problem

Yellowstone National Park is full of life, and scientists are intrigued with the abundance of interactions within the ecosystem. They are always thinking of new research questions, and they continue to observe plants and animals in the park. Your task is to design and build a tool that scientists can use to study and make observations about plants or animals at Yellowstone. Your tool can either be something that scientists wear or use to help them do their work.

Constraints: You may use only the materials provided to you.

Criteria: You must be able to wear the tool you created or demonstrate how it is used. It must include at least two key features for observation, and you must be able to explain how it will be useful to scientists.

Research and Brainstorm

What types of animals and plants may scientists be interested in observing at Yellowstone? What are some ways that animals coexist within the Yellowstone ecosystem?

Design and Build

Sketch two or more designs for your observation tool. Label the parts and materials. Choose the design you think will work best. Then, build your prototype. You can test the gear or tool as you build and make adjustments to your design.

Test and Improve

Share your tool with others. Explain how it can be used to study animals or plants at Yellowstone. Demonstrate how it is used by having a team member try it on or by performing a task with it. What about your design is working well? How can you improve the look or function of your design? Modify your design and rebuild as needed. Reassess how well it meets the criteria.

Reflect and Share

What about this challenge did you find most interesting? How could you modify or add to your design to make it useful for other environments? What would you want scientists to know about the design before using it in the field?

Glossary

carrion—the decaying flesh of dead animals

climate change—important and long-lasting changes in Earth's climate and weather patterns

coexist—to live in peace with one another

conservation—planned management of a natural resource or species to prevent exploitation and promote protection

ecosystems—groups of living and nonliving things that make up an environment and affect one another

enzymes—special proteins produced by cells that start or speed up chemical reactions

food chain—a series of organisms in which each is dependent on the next as a food source

geysers—pools of naturally hot water that occasionally boil, shooting water and steam into the air

hydropower—electricity created by fast-moving water, such as from a dam on a river

microbes—microscopic organisms, such as bacteria

migrate—to move from one area to another according to the seasons

organic—of, relating to, or obtained from living things

overgrazing—excessive eating by animals that causes damage to land

park rangers—people in charge of managing and protecting parts of national parks

pests—destructive insects or other animals that are harmful to plants, animals, and humans

photosynthesis—a chemical process in which plants make their food using water, sunlight, and carbon dioxide

predators—animals that get food by hunting other animals

sedges—grasslike plants that are a source of food for animals

snowmelt—water produced by melting snow

suckers—freshwater fish that suck up food from the bottoms of bodies of water

thermophiles—microscopic organisms that grow best at hot temperatures

Index

algae, 9–10, 17
archaea, 17
bacteria, 12–13, 17–18
bison, 4, 9–11, 14–15, 23
Castle Geyser, 18–19
climate change, 5, 20, 23–26
consumers, 7–8, 10–12
decomposers, 7–8, 10, 12–13
dung beetles, 15
ecosystem, 5–8, 11, 14, 20, 24–26
elk, 4, 9, 11, 14–15, 23
eukarya, 17
flood, 22–23, 26
fungi, 7, 12, 26

geysers, 4, 6, 16–19, 26
gray wolves, 4, 11, 14
Hidatsa, 27
hot springs, 4, 16–17
Moran, Thomas, 6
Old Faithful, 18–19
park rangers, 25
producers, 7–12
Pump Geyser, 18
ravens, 14
scavengers, 13
thermophiles, 4, 16–18
Upper Geyser Basin, 18
whitebark pine, 20
wildfire, 20, 26

bison along the Yellowstone River

CAREER ADVICE
from Smithsonian

Do you want to study ecosystems?

Here are some tips to keep in mind for the future.

"Every living thing can be a teacher; to be students takes curiosity and a willingness to learn. Take classes in STEAM that interest you and remember that the world is your classroom, too. Practice embracing your curiosity and close-looking at the environment around you, wherever you are. Patience, observation, and dedication are key."

– Alia Payne, Ocean Education Specialist, Smithsonian National Museum of Natural History

"Scientists will make many observations while learning about the environment that they are studying. There is no better place to get this practice than to make observations in your local environment! Take the time to explore the forests, streams, lakes, coastlines, deserts, or mountains that are around you and make close observations of nature. Even if it is your own backyard, you will be surprised how many things you will see in a small area when you look closely. There was always something that surprised me in my local area when I was growing up."

– Bayley McKeon, Ocean Education Specialist, Smithsonian National Museum of Natural History